Kingsley Adimabua, G. A. Awemu, A. U. D. Aniekan, A. M. Odiegwu

Antimicrobial and Phytochemical Studies on Extracts of Bucholzia coriacea Seeds

AF400503

Kingsley Adimabua, G. A. Awemu, A. U. D. Aniekan, A. M. Odiegwu

Antimicrobial and Phytochemical Studies on Extracts of Bucholzia coriacea Seeds

GRIN Verlag

1. Auflage 2013
Copyright © 2013 GRIN Verlag GmbH
http://www.grin.com
Druck und Bindung: Books on Demand GmbH, Norderstedt Germany
ISBN 978-3-656-36512-9

ANTIMICROBIAL AND PHYTOCHEMICAL STUDIES ON EXTRACTS OF *BUCHOLZIA CORIACEA* SEEDS

G. A. Awemu,[1] A.U.D. Aniekan,[1] A.M. Odiegwu[2], K.O. Adimabua,[3]

[1]Department of Pharmaceutical and Medicinal Chemistry, Madonna University, Elele, Nigeria

[2]Department of Nursing Science, Madonna University, Elele, Nigeria

[3]Department of Public Health, Madonna University, Elele, Nigeria

Abstract

The in vitro antimicrobial activity of crude methanol and aqueous extracts of the seeds of *Bucholzia coriacea* were investigated. The extracts exhibited antimicrobial activities against Escherichia coli, klebsiella pneumonia, Bacillus subtilis, Staphylococcus aureus, Salmonella typhii, Bacillus cereus and Pseudomonas aeruginosa. The minimum inhibitory concentration (MIC) of the ethanol extract was between $0.50 - 6.00$ mgml^{-1} while the minimum bactericidal concentration ranged from $2.0 - 10.0$. The methanol and water extracts exhibited antifungal activity against Candida albicans and Aspergillus niger with zones of inhibition of 7.50 and 2.80mm for Candida albicans; and 6.0 and 2.0 for Aspergillus niger. Phytochemical screening revealed the presence of tannins, saponins, terpenoids, cardiac glycosides and alkaloids in the ethanolic and water extracts. The ability of the ethanol extract of Bucholzia coriacea seeds to inhibit the growth of bacteria and fungi is an indication of its broad spectrum antimicrobial potential which justifies its utilization in traditional medicine in treatment of infections.

Introduction

Antimicrobial agents are among the most commonly used and misused classes of drugs leading to the emergence of drug resistance fuelling an ever increasing need for new drugs (Abdullahi et al, 2010; Chambers, 2006). To overcome the problem of antibiotic resistance, medicinal plants have been extensively studied as alternative treatments for infective diseases, as they inhibit the growth of pathogens or kill them and have no or least toxicity to host cells (G. Sahgal et al, 2009).

Bucholzia coriacea belongs to the family Capparacea and is widely distributed in the rain forests of Ivory Coast, Cameroon, Liberia, Nigeria and Gabon (Ajaiyeoba et al, 2003). It

is a medium sized evergreen tree or shrub with large, glossy, leathery leaves that are arranged spirally and clustered with conspicuous cream white flowers in racemates at the end of the branches. Its bark is smooth, dark brown or dark green while the slashes are deep red (Dela Veau , et al, 1973). The seed is commonly called wonderful kola or elephant kola in Nigeria. The leaves and stem bark in various formulations exhibit anthelminthic, antimicrobial, antifungal and cytotoxic effects on microorganisms (Ajaiyeoba et al, 2003). Despite the various applications of the plant in ethnomedicine, very little scientific evaluation of the seed extracts have been carried out. The present study was therefore undertaken to evaluate the antimicrobial effects and phytochemical profile of the methanolic and aqueous extracts of *Bucholzia coriacea* seeds.

MATERIALS AND METHODS

PLANT MATERIAL AND EXTRACTION

Bucholzia coriacea pods were harvested in the forest at Elele, Rivers State of Nigeria and its identity authenticated by a taxonomist at the Department of Pharmacognosy, Faculty of Pharmacy, Madonna University, Elele, where a voucher specimen was deposited. The fresh seeds were removed from their pods, sliced into tiny particles to facilitate drying and shade dried for fourteen days. The dried seeds were pulverized using an electric mill. A 250 g of the powder was extracted by maceration in ethanol for forty eight hours. The extract was filtered using Whatman no. 1 filter paper and concentrated using a rotary evaporator. The resultant extract was stored in a refrigerator at 4°C. For the aqueous extract, 250 g of the powered seed was extracted through cold maceration with water as the solvent for forty

eight hours. The resultant extract was concentrated with a rotary evaporator and stored in a refrigerator at 4^0C.

PHYTOCHEMICAL TESTS

Both the aqueous and methanolic extracts were independently screened for the presence of tannins, saponins, terpenes, cardiac glycosides and alkaloids according to the standard methods (Trease and Evans, 1983).

MICROORGANISMS

Staphylococcus aureus, Bacillus subtilis, Escherichia coli, Klebsiella pneumonia, Salmonella typhii, Bacillus subtilis, Pseudomonas aeruginosa and the fungi, candida albicans and Aspergillus niger were used in the study. They were all obtained from the Federal Medical Center(FMC), Owerri, Imo State of Nigeria as clinical isolates and maintained in nutrient broth media at 37°C in the Department of Pharmaceutics and Pharmaceutical Microbiology, Madonna University, Elele.

ANTIBACTERIAL ACTIVITY

The antibacterial activity of the extracts were determined in accordance with the agar well diffusion method(Irobi et al, 1994 and Igbinosa et al, 2009). The bacterial isolates were first grown in nutrient broth for 18H before use and standardized to 0.5 McFarland standard (10^6 cfuml^{-1}). Two hundred microlitre of the standardized cell suspension were spread on a Mueller – Hinton agar (Oxoid). Wells were then bored into the agar using a sterile 6 mm diameter cork borer. Approximately 100 µl of the crude extract at 10mgml^{-1} were introduced into the wells, allowed to stand at room temperature for about 2 hours and then

incubated at 37°C. the plate were observed for zones of inhibition after 24 hours. The effects were compared with those of ciprofloxacin at 1mgml^{-1}.

Anti fungal activity

The fungal isolates were allowed to grow on a sabourand dextrose agar (SDA) (Oxoid) at 25°C until they sporulated. The fungal spores were harvested after sporulation by pouring a mixture of sterile glycerol and distilled water to the surface of the plate and later scraped the spores with a sterile glass rod.

Minimum inhibitory concentration (MIC)

The MIC was determined for the microorganisms that showed sensitivity to the test extracts. The broth dilution method was used for MIC determination according to the method of Vollekova et al (2001). The prepared broth was poured in then test tubes and inoculated with 2 ml of the sensitive microorganisms. Several dilutions of the extract and standard were prepared and 0.1 ml of each was transferred in each test tube and labeled. The test tubes were inoculated for 24 hrs at 37°C and the least concentration in which no growth of the microorganisms (absence of turbidity0 was noted as the minimum inhibitory concentration.

Minimum bactericidal concentration (MBC)

The method of Spencer and Spencer (2004) was used for the MBC determination. Samples were taken from the plates with no visible growth in the MIC assay and subcultured on freshly prepared nutrient agar plates and SDA plates, and later incubated at 37°C for 48 hrs and 25°C for 72 hrs for bacteria and fungi respectively. The MBC was taken as the concentration of the extract that did not show any growth on a new set of agar plates.

Results and discussion

The phytochemical screening of the methanol and water extracts revealed the presence of alkaloids, anthraquinones, cardiac glycosides, flavonoids, reducing sugars, saponins, tannins and terpenoids. These classes of compounds are known to show curative activity against several pathogens and therefore could explain its use traditionally for the treatment of a wide array of illnesses (Hassan et al, 2004; Usman et al, 2005 and Usman and Osuji, 2007). The ethanol extract of Bucholza coriacea seed showed varying degree of antibacterial activities against test microorganisms (Table 1). The antibacterial activities of the extract compared favourably with that of the standard antibiotic ciprofloxacin, and the broad spectrum of its activities were independent of gram reaction (Igbinosa et al 2009).

The inhibition zone for Klebsiella spp. was the lowest (6.00 ± 0.12 mm) while that for E. coli was the highest (15.40 ± 0.24mm) in all the test bacteria studied. The methanol extract showed greater antimicrobial activity than the water extract. This may be attributed to the presence of soluble phenolic and polyphenolic compounds (Kowalski and Kedzia, 2007). The water extract showed low antibacterial activity with inhibition zones ranging between 2.80 and 4.00 for different bacteria tested.

The Minimum Inhibitory Concentration (MIC) of the ethanol extract for different organisms ranged between 0.50 to 6.00 mgml^{-1}. Also the MIC of Ciprofloxacine ranged between 0.065 and 0.625 mgml^{-1}. The Minimum Bactericidal Concentration (MBC) of the extract for different bacteria ranged between 2.0 and 10 mgml^{-1}(Table 2). The water extract was not active against any of the organism at 10 mgml^{-1}. At a concentration of 7.5 mgml^{-1}, the methanol extract exhibited antifungal activity with inhibition zone diameter of 7.50 mm. The antifungal activity is quite significant and is higher than that for the water extract at 10 mgml^{-1} with an inhibition zone diameter of 2.80 mm.

The inhibitory effect of the ethanolic extract of B. coriacea seed against pathogenic

bacterial strains can introduce the plant as a potential candidate for drug development for

the treatment of ailments caused by these pathogens. The non activity of the water extract

against most bacterial strains investigated in this study is in agreement with previous works

which show that aqueous extracts of plants generally show little or no bactericidal activity (

Koduru et al, 2006 and Igbinosa et al, 2009).

Table 1. antibacterial and antifungal profile of Bucholzia coriacea seed extracts

Test Bacteria	Zone of inhibition mm (mean ± SEM)		
	Ethanol(10 mgml^{-1})	Water (10 mgml^{-1})	*Reference drug (1mgml^{-1})
Escherichia coli	15.4 ± 0.24	4.0 ± 0.1	28.0 ± 0.3
Klebsiella pneumonia	6.0 ± 0.12	3.8 ± 0.09	15.0 ± 0.61
Bacillus subtilis	7.8 ± 0.12	3.4 ± 0.1	20.0 ± 1.0
Staphylococcus aureus	13.4 ± 0.36	3.8 ± 0.18	25.0 ± 1.2
Salmonella typhii	8.2 ± 0.20	3.5 ±0.2	20.0 ± 0.2
Bacillus cereus	7.8 ± 0.3	3.1 ± 0.1	22.0 ± 1.1
Pseudomonas aeruginosa	7.0 ± 0.1	3.3 ± 0.2	20.0 ± 0.3
Candida albicans	7.5 ± 0.08	2.8 ± 0.1	19.0 ± 0.8
Aspergillus niger	8.0 ± 0.1	2.0 ± 0.2	16.0 ± 0.5

*Miconazole (1mgml^{-1}) was used for candida albicans and Aspergillus niger while ciprofloxacine (1mgml^{-1}) was used for the other test bacteria.

Table2. The MIC and MBC regimes of the ethanol extracts of Bucholzia coriacea see

Test Bacteria	Ethanol (mgml^{-1})		Ciprofloxacin (mgml^{-1})
	MIC	MBC	MIC
Escherichia coli	6.00	10.00	0.065
Klepsiella pneumonia	2.00	5.00	0.500
Bacillus subtilis	0.50	2.00	0.625
Staphylococcus aureus	5.00	2.00	0.065
Salmonella typhii	2.50	4.00	0.500
Bacillus cereus	0.50	2.00	0.500
Pseudomonas aeruginosa	4.00	2.00	0.500

REFERENCES

1. M.I. Abdullahi, I. Lliya, A. K. Haruna, M. I. Sule, A. M. Musa, M.S. Abdullahi (2010). Preliminary phytochemical and antimicrobial investigations of leaf extracts of Ochna schweinfurthiana (Ochnaceae). African Journal of Pharmacy and Pharmacology. Vol. 4(2) pp. 083.

2. F.H. Chambers (2006). General principles of antimicrobial therapy. In: The pharmacological basis of therapeutics, Bruton, L.L., Lazo, J.S. and Parker, L.L. (Eds.). McGraw Hil publishers, New York, p. 1095.

3. G. , Sahgal, S., Ramanathan; S., Sasidharan; M.N. Mordi; S, Ismail and S.M. Mansor (2009). Phytochemical and antimicrobial activity of Swretenia mahagoni crude methanolic seed extract. Tropical Biomedicine 26(3): 274-276.

4. P. Delaveau., B. Koudogbo., and J.L. Pousset (1973). Alkaloids chez les capparidaceae. Phytochem, 12: 2893 – 2895.

5. G.E. Trease and W.C. Evans (1983). Textbook of Pharmacognosy, 12th edition. Balliers Tindall, London, pp. 343 – 384.

6. O.N. Irobi, M. Moo-Young, W.A. Anderson, and S.o. Daramola (1994). Antimicrobial activity of the bark of Bridelia ferugunea (Euphorbiaceae0. Intern. J. pharmacog. 34: 7 – 90.

7. O.O. Igbinosa, E.O. Igbinosa and O.A. Aiyegoro (2009). Antimicrobial activity and phytochemical screening of stem bark extracts from Jatropha curcas (Linn.). african Journal of Pharmacy and Pharmacology, 3(2): 58-62.

8. R. Kowalski and B. Kedzia (2007). Antibacterial activity of Silphium perfoliatum extracts. Pharm. Biol. 45: 495 – 500.

9. S. Koduru, D.S. Grierson and A.J. Afolayan (2006). Antimicrobial activity of Solanum acvuleastrum (Solanaceae). Pharmacol. Biol. 44: 284 – 286.

10. E.O Ajaiyeoba, P.A. Onocha, S.O. Nwodo and W. Sama (2003). Antimicrobial and cytotoxicity evaluation of Bucholzia coriacea stem bark. Fititerapia 74(7-8): 706 - 709.